MSM:

On Our Way Back TO Health With

SULFUR!

A Health Learning Handbook

By Beth M. Ley, Ph.D.

BL Publications
Hanover, MN

Table of Contents

BL Publications, 1-877-BOOKS11
www.blpublications.com

Printed in the United States of America

Credits:
Research and Technical Assistance: Sid Shastri, Keith Morey
Editing: Kim DaLaura, Harold Dimond

Introduction

Health Learning Handbooks are designed to provide interesting and useful information to improve one's health and well-being through what the body needs to obtain and maintain good health.

Good health should not be thought of as the absence of disease. We should avoid this negative disease-orientated thinking and concentrate on what we must to do to remain healthy. Health results from supplying what is essential to the body on a daily basis, while disease results from living without what the body needs. We are responsible for our own health and should take control of it before diseases does.

Our health depends on education. Education is power.

My dear friend, I pray that you may be in health; I know that it is well with your soul. **3 John: 2**

I usually do not write about myself in my books, but I want to share my experience with MSM. I do experiment with almost all of the substances I write about. If I don't believe that they work and have value, I wouldn't want to spend the time writing about them.The first time I tried MSM, I didn't see any effect so I didn't give it much thought after that. People continued to ask me to do a book on the subject. I looked into it further and tried it again. I was told that I didn't take a high enough dosage the first time, so I will discuss dosage thoroughly later in the book. This time, in just three days, I noticed a profound difference in the clarity of my lungs. I could hardly believe it. Of all the products I have tried, I have never experienced such a strong effect. Sublingual DHEA has a powerful effect, but only when used to treat symptoms during an attack. MSM brought about a clarity all day long, helping to ward off attacks.

I started my investigation, spoke with people and found out some incredible things. I hope you enjoy the book and I wish you good health.

MSM Background

MSM stands for Methyl-Sulfonyl-Methane (Dimethyl Sulfone. This is a nonmetallic sulfur compound which occurs widely in nature. Sulfur is a yellow material which has a large number of physical forms, crystalline as well as liquid. It is one of the most common substances found in the body. Sulfur plays an essential role in human nutrition, a fact commonly overlooked.

MSM is a derivative of DMSO (dimethy-sulfoxide). DMS0 is widely used to treat animals such as horses to reduce inflammation in joints and injured areas. The side effects of smell and impurities limit the widespread use of DMS0 in humans.

MSM is a white, odorless crystalline material resembling sugar. The sulfur component of MSM is 34%, by weight, making MSM one of nature's richest sources of sulfur. MSM tastes slightly bitter and mixes easily into water or juice as the solubility of MSM is very high. MSM is as basic to life as water and salt. It is completely non-toxic - as safe as water!

MSM belongs in the same chemical family which includes oxygen. For organisms living in an environment without oxygen, sulfur often replaces oxygen as the source of chemical energy that drives life.

Made in the Clouds!

Sulfur is present in the cells of all animals and plants. It is needed by all plants and animals, but in order to use it, it must in a bioavailable form.

Simple microscopic plant and animal life (plankton, algae, etc.) assimilate inorganic sulfur and release organic sulfur compounds called dimethylsulfonium salts. These salts are transformed in ocean water into the very volatile compound dimethylsulfide (DMS). This escapes from the ocean as a gas and rises into the upper atmosphere. There, in the encouragement of ozone and high energy ultraviolet light, DMS is oxidized into dimethyl sulfoxide (DMSO). DMSO is oxidized to methyl-sulfonyl-methane (MSM). These two compounds dissolve in atmospheric moisture, are caught up in the clouds and fall to the earth when it rains or snows. (Thus, the cover photo for the book)

Unlike the DMS, both DMSO and MSM are very soluble in water. Plants absorb DMSO and MSM from rain water, concentrating them up to one hundred fold. MSM converts into methionine and cysteine which are then incorporated into the plant structure. Through the processes of plant metabolism, MSM, and other sulfur compounds are transformed, eliminated, and returned to the sea. Then process is then repeated.

The methionine and cysteine in plants provide us a rich source of organic sulfur. These two amino acids, along with others, combine to form proteins. a plentiful vegetative food source of organic sulfur. These two amino acids, along with the others form protein. Animal protein is another source of sulfur.

MSM and its related compounds, DMSO, and DMS, are the sources for 85% of the sulfur found in living organisms. (Lovelock)

The flora of the world utilize sulfur to form various penicillins. Sulfur is responsible for the characteristic odors of garlic, onion, mustard, horseradish, etc. Many other peculiar odors in nature are due to sulfur(s) compounds. The odor of burned hair or wool indicates its high sulfur content and the strange odor of urine following the consumption of asparagus are additional examples.

Also interesting is that the curliness of hair depends on the sulfur-to-sulfur bonds of cystine. Hair straighteners are designed to disrupt S-S bonds.

Every wonder why snow is white?

Water, containing sulfur and/or other minerals appears white (like snow, or waterfall or glacial ice) when frozen. When you freeze distilled or even tap water, it freezes clear.

We have polluted our environment such that in most areas that it is no longer safe to drink well water, river or mountain stream water like we did at one time. We now drink purified water no longer containing minerals necessary for good health.

When you freeze water containing MSM, it appears white. When melted it appears clear, as when snow melts.

Why Do We Need Sulfur?

Sulfur is found naturally in the human body. **Sulfur is stored in every cell of the body.** The highest concentrations are found in the joints, hair, skin, and nails. Excess sulfur is excreted in the urine (about four to 11 mg. MSM per day) and the feces.

Sulfur also occurs in the blood and in the other organs as well. It has been detected in normal human urine (Williams). The natural level of MSM in the circulatory system of an adult human male is about 0.2 ppm (Jacob).

As you can see by examining the following chart, sulfur is the eighth most abundant substance in the body by weight. (Brown) If you eliminate water and gas, sulfur is the surprising third most abundant substance!

Substance Composition in Man by Weight

	Substance	Weight
	Water	45,000 g
	Carbon	16,000 g
	Oxygen (Nonaqueous)	2,900 g
	Hydrogen (Nonaqueous)	2,000 g
	Nitrogen	1,800 g
1.	Calcium	1,100 g
2.	Phosphorus	600 g
3.	**Sulfur**	**140 g**
	Potassium	140 g
	Sodium	100 g
	Chlorine	95 g
	Magnesium	19 g
	Silicon	18 g
	Iron	4.2 g
	Zinc	2.3 g
	Vitamin C	1.2 - 2.0 g

In mammals the concentration of MSM in the body's various storehouses decreases with age, possibly as a result of changing diet or body metabolism. Some research suggests that there is a minimum concentration of MSM that must be maintained in the body to preserve the normal function and structure. (Jacob)

The body uses MSM to continually create new healthy cells to replace old ones. The body uses vitamins and amino acids with MSM in this process. If the body isn't receiving the proper nutrition including MSM, the body instead will produce weak dysfunctional cells.

Sulfur works with thiamine, pantothenic acid, biotin, and lipoic acid, which contain elemental sulfur, are needed for metabolic processes. They also contribute to nerve health. Sulfur plays a part in tissue respiration, the process whereby oxygen and other substances are used to build cells and release energy. It works with the liver to secrete bile. Sulfur also helps to maintain overall body balance.

Sulfur is an essential mineral for the human body and plays many roles. Here is a rough overview of three of its major responsibilities in the body.

1, Dehydration and Detoxification

Sulfur is responsible for the ionic exchange within the sodium-potassium pump in the cell. Sulfur is needed to maintain cell membrane permeability. This is important to ensure that nutrients are delivered into the cell and toxins and waste products can exit the cell.

2. Energy

Sulfur is a component of insulin, the very important hormone that regulates the uptake of glucose by cells for use as energy. Sulfur is also needed for thiamine and biotin which are also needed for normal carbohydrate metabolism.

3. Structure and Function

Sulfur is a required element found in the proteins of most body tissues: skin, blood vessels, organs, hair and nails. Sulfur makes up the flexible S-S (disulfide) bonds within proteins which provide elasticity and flexibility for movement. (Kim) Sulfur is also involved in the repair and healing of these tissues damaged by injury, free radical attack, aging, etc. All cells in the body are in a state of continual regeneration, each at a different rate.

Rate of cellular renewal: *(Fastest to slowest)*

1. Epithelial cells - make up the tissues that cover the outside of the body (skin) and line the digestive, respiratory and urinary tracts.

2. Connective tissue - includes fats cells, mucopolysaccharides and fibrous cartilage which maintains elasticity, flexibility, and holds the body together.

3. Muscle, nervous and bone cells - Nerve and bone cells may take up to seven years to replace themselves.

Are You Sulfur Deficient?

Sulfur deficiencies are associated with:

Slow wound healing
Scar tissue
Brittle nails
Brittle hair
Gastrointestinal problems
Regulation of inflammation
Lung disfunction
Immune disfunction
Arthritis
Acne
Rashes
Depression
Memory loss

Supplementing MSM assures that you have an adequate organic source for the body to use for what ever it needs to. Remember the body is in a constant state of self repair, but if we do not have all the necessary "parts", the end product is going to be less than perfect.

Scientists around the world are investigating the relationship between MSM and arthritis (Sharif, Malfait Rizzo), Alzheimer's disease, allergies and asthma (Joris), dermatological problems (Agostini), periodontal conditions (Miyazaki) and even cancer.

Sources of Sulfur

Before the existence of modern diets containing high amounts of processed foods, our foods were an important source of MSM and sulfur. MSM was found naturally and all forms of life were able to obtain it for use.

Sulfur is a natural component of many foods. The level of sulfur in soil varies a great deal and many areas are grossly deficient in sulfur; therefore, sulfur content in plant foods will vary and may also be deficient.

Foods containing sulfur include:

Fresh Fruits

Fresh Vegetables, (0.6 ppm in tomatoes to 0.11 ppm in corn). Red hot peppers, cabbage, brussels sprouts and horseradish contain higher amounts

Dried Beans, including Soybeans (.38% sulfur)

Fish and Seafood

Meat (1.27% sulfur)

Eggs (whites contain 1.62% sulfur)

Milk (.8% sulfur), and Cheese

Tea

Coffee

Chocolate

Garlic and Onions

Horsetail (herb)

Wheat germ

Amino acids: L-cysteine, L-cystine, L-methionine.

Lack of Sulfur in Soil

Air and water pollution, over-farming and polluted soil, artificial irrigation practices, deforestation and etc., contribute to mineral depleted soil. Mineral depleted soil leads to mineral depleted food and mineral depleted people.

Commercial fertilizers seldom restore trace elements to the soil. Plants depend on the soil for sulfur in the form of the sulfate ion. This is taken into the plant, where enzymes convert the sulfate into the many organic sulfur compounds which plants and animals need.

Many foods are now inadequate sources of MSM and sulfur because of losses during food processing. Today, just as many other once abundant minerals are now lacking, it is not as easy to obtain adequate dietary sulfur.

There is no Recommended Dietary Allowance for sulfur. It is assumed that a person's sulfur requirement is met with an adequate protein intake. Vegetarians are at risk of a sulfur deficiency, especially lacto-ovo vegetarians.

MSM is a dietary supplement. There is limited human clinical data regarding dosage, therefore, recommendations tends to vary. Optimum dosage depends upon several factors: age, sex, body mass and the amount of MSM in the blood.

According to the pioneers of MSM-related research, Dr. Stanley Jacob and Dr Robert Herschler, at Oregon Health Sciences University in Portland, improvement in several health disorders are seen in the range of 250 to 750 mg. per day. Earl Mindell, Ph. D., MSM recommends a daily intake of 2 to 6 grams. (Mindell)

A Body Made of Sulfur

Sulfur is an important element in more than 150 compounds in the body including tissues, enzymes, hormones, antibodies, and antioxidants. Following is a summary of a few of the many sulfur-containing compounds:

Sulfur Amino Acids

Sulfur is present in amino acids such as methionine, taurine, cysteine and cystine. Except for methionine, all the sulfur-rich amino acids can be synthesized by the body from methionine and sulfur. In most instances, the major sources of animal dietary sulfur are the two amino acids, methionine and cysteine. From these the body builds the essential compounds coenzyme A, heparin, glutathione, lipoic acid and biotin.

Daily requirements are given as 10 mg a day per kilogram of body weight for all the sulfur amino acids, including methionine.

Sulfur-containing amino acids help protect against the effects of radiation and heavy metal chelators.

Methionine, an essential amino acid, is an important source of sulfur. It is required for DNA-RNA structure, collagen, and each cell's protein synthesis function. Methionine is involved in the formation of lecithin, which helps breakdown down fats in the liver and bloodstream.

Methionine is a powerful detoxifying agent. It is able to detoxify histamine. Through chelation, methionine can eliminate toxic metals such as lead, cadmium, and mercury. Supplementation is required to obtain an effect upon the body's toxic metal burden. (Schauss)

Methionine is an effective scavenger of free radicals. It can deactivate super oxide radicals derived from alcohol. Methionine also aids in the maintenance of glutathione peroxidase, the powerful enzyme antioxidant.

Methionine is an anti-oxidant. However, its derivative homocysteine is a powerful oxidant. Elevated levels of homocysteine are associated with an increase risk of heart disease. Adequate levels of vitamin B-6 allow this to be reconverted into an antioxidant substance, cystathione.

Methionine is also essential to absorb, transport and utilize selenium, an important antioxidant trace mineral.

Methionine is a precursor to the amino acids cysteine and cystine. Methionine is found primarily in the following foods: beef, chicken, fish, pork, soybeans, egg, cottage cheese, liver, sardines, yogurt, pumpkin seeds, sesame seeds and lentils.

Deficiency symptoms include poor skin tone, hair loss, toxic waste buildup, and liver malfunction. Methionine has a particularly distinctive sulphurous odor which most people find unpleasant.

Cystine is the stable form of the sulphur-rich amino acid **Cysteine**. The body is capable of converting one to the other as required. In metabolic terms they can be thought of as the same.

Methionine and cysteine are utilized in the formation of a number of essential compounds, such as coenzyme A, heparin, biotin, lipoic acid and glutathione. Cysteine is a vital component of insulin and the glucose tolerance factor.

Cystine is found in our hair, keratin, insulin, digestive enzymes and immunuglubulins. The flexibility of the skin, as well as the texture is influenced by cysteine as it has the ability to slow abnormal cross-linkages in collagen, the connective tissue protein. Cysteine will convert cystine in the absence of vitamin C.

Cystine deactivates free radicals (especially those induced by alcohol and cigarette smoking), and to strengthen immunity (helps repair DNA-RNA cellular components, etc.).

Deficiency symptoms of cystine include poor skin tone, hair loss, toxic waste buildup, and liver malfunction.

Cysteine is more soluble than cystine. Both have a strong sulfur odor and taste. The best way to supplement cystine is in the form N-acetylcysteine (NAC).

Taurine, a sulfur amino acid derivative, is manufactured in the body. It is found in animal protein, but not in vegetable protein. Its synthesis in humans is from the amino acids methionine and cysteine, primarily in the liver with the assistance of vitamin B6.

Dietary intake is thought to be more necessary in women, since the female hormone estradiol depresses the formation of taurine in the liver. Any additional estradiol in the form of medication would increase this inhibition.

Taurine is a component of bile, needed to digest fats, regulate cholesterol levels and absorb fat-soluble vitamins.

Taurine is sometimes called the "brain amino acid" as it works with choline to help maintain our neurotransmitters. Taurine is found in the developing brain in concentrations up to four times that found in the adult brain. Since taurine acts as a suppressor of neuronal activity in the developing brain, during the phase when other regulatory systems are not fully developed, it is thought that deficiency of taurine, at this stage, might contribute towards the development of epilepsy. Taurine has been shown in human trials to have an anticonvulsive effect. It may help normalize the imbalance of other amino acids seen in epilepsy.

Taurine contributes to the health of the heart and the control of blood pressure. It regulates calcium and potassium levels in heart muscle. This is especially important when dieting. Any strict weightloss program should include the addition of sulfur-bearing amino acids, such as methionine and cysteine. This ensures adequate availability of taurine to protect the heart muscle against loss of calcium and potassium.

Taurine influences blood sugar levels, similar to that of insulin. Heart conditions, such as myocardial infarction, or skeletal damage, physical or emotional stress, and diseases involving platelet or leucocyte haemolysis, are all potential causes of increased taurine excretion in the urine.

High alcohol consumption, use of aspirin, and zinc deficiencies are associated with elevated urinary excretion of taurine. Taurine deficiencies are associated with depression, hypertension, hypothyroidism, and kidney failure, and supplementation (500 to 5,000 mg) is recommended. Large oral doses of taurine may stimulate production of growth hormone.

Additional Sulfur-containing Compounds:

Glutathione: This amino acid is derived from the sulfur-bearing amino acid cysteine plus glycine and glutamic acid. Glutathione is produced mainly in the liver. It is found in virtually all living cells.

Glutathione is one of the key antioxidants in the body and responsible for protecting the body waste disposal. Glutathione plays a role in cellular repair after a stroke, fighting cancer, stabilizing blood sugar and preventing oxidation of LDL cholesterol which damages the arteries. It's also crucial in protecting the lymphatic and digestive systems from unstable fats and oils. When glutathione levels drop, the burden of toxic stress goes up.

Our level of glutathione drops as we get older. Levels can also be depleted by excessive intake of polyunsaturated and partially hydrogenated vegetable oils, overexposure to toxic substances such as pesticides, and by pharmaceutical drugs that stress the liver.

Since glutathione works with vitamins C and E, a deficiency of these vitamins can diminish its effectiveness.

Glucosamine is the building block for ligaments, tendons, fluid in the joints, digestive and respiratory tract membranes, heart valves, eyes, nails, skin and bone. It gives cartilage its strength, structure and resiliency.

It is produced in the body by sulfation of glucose and glutamine. Sulfur bonds are an important structural components of all connective tissue. Connective tissue is found everywhere in the body. It supports and connects our internal organs, forms the walls of blood vessels, and joins muscles to bones. One component of connective tissue is collagen, which holds water and gives connective tissue its flexibility. Proteoglycans are another component of connective tissue, and they are the basic substance of joint cartilage.

Glucosamine supplementation has no known side effects, contraindication or drug interactions.

Homocysteine is a natural amino acid created as a by-product of methionine metabolism. Homocysteine is toxic but in a healthy body it is quickly transformed into harmless substances before it builds up and causes harm.

Homocysteine binds with LDL cholesterol. This causes damage to the arterial wall, by causing a buildup of plaque, and, eventually, atherosclerosis. High levels of homocysteine largely result largely from dietary inadequacies of B-6,

Homocysteine in cells and tissues

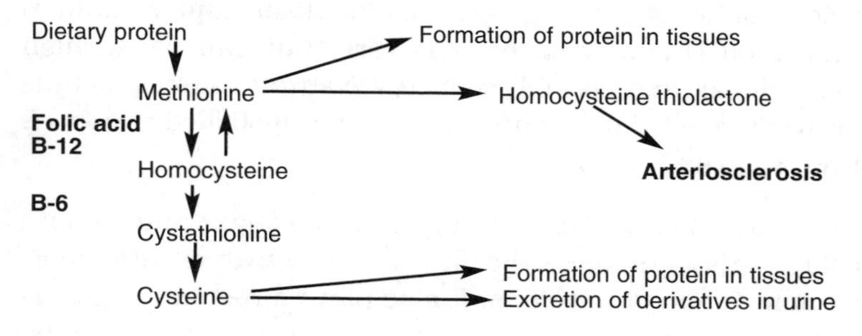

B-12 and Folic acid.

Homocysteine thiolactone is a highly reactive form of homocysteine which causes LDL cholesterol to aggregate or plaque in the bloodstream. Foam cells also release homocysteine thiolactone into the surrounding cells creating free radicals which damage the lining cells or the artery wall. This is the beginning of arteriosclerosis and increased risk for blood clots, etc. (McCully)

Studies show that men and older women tend to have higher levels of homocysteine. Studies also show that the higher the level of homocysteine, the higher the risk of heart problems. The higher your level of folic acid, the lower the level of homocysteine. Therefore, increasing folic acid levels can prevent a buildup of homocysteine in the blood.

Folic acid, B-12, B-6, and Betaine (also known as Trimethylglycine or TMG) can help reduce elevated levels.

Individuals eating a high protein diet, on certain medications and a increased toxic load (smokers, drinkers, etc) have an increased need for these important nutrients. (McCully)

Thiamine (Vitamin B-1) helps convert carbohydrates to energy. B-1 is necessary for proper functioning of the heart and nervous system. Thiamine helps build the blood, prevents fluid retention, prevents constipation and maintains muscle tone. Thiamine also serves an an antioxidant. Natural sources include whole grains and cereals, yeast, liver, pork, poultry, lean meat, eggs, and legumes.

Biotin is a B vitamin needed for cell growth and fatty acid production, hair growth, metabolism, and vitamin B utilization. Biotin deficiency is rare, but can cause high blood sugar, and possibly diabetes. Natural sources include brewer's yeast, fruits, nuts, soybeans, unpolished rice, beef, liver, egg yolk, and milk.

Alpha Lipoic Acid is a potent antioxidant and vitamin cofactor that enhances the antioxidant activity of vitamin E, vitamin C and glutathione. It also plays a role in the generation of energy and in glucose balance. In Europe, alpha lipoic acid has been used for nearly 50 years to treat diabetes and as an antioxidant to treat and prevent polyneuropathy, cataracts and macular degeneration.

Alpha-lipoic acid helps the liver detoxify heavy metals, fight alcohol-induced liver disease and treats viral hepatitis.

Coenzyme A is derived from pantothenic acid, (B-5). It is important in the synthesis of fatty acids, cholesterol and derivatives of cholesterol (bile, vitamin D, and steroid hormones). Co-enzyme A is involved in the production of red blood cells and the neurotransmitter acetylcholine. Coenzyme A deficiency is relatively rare.

Collagen, the single most common protein in the body helps form bones, tendons, and connective tissue. It helps bind cells and body tissues together.

Keratin, a fibrous protein, is the primary component of skin, hair, nails and tooth enamel. Keratin constitutes 98% of nails.

Fibrinogen is a necessary component of blood as a clotting agent, but when there is too much of it, your blood becomes too sticky making plaque deposits worse which restricts blood flow.

Insulin is the very important hormone produced in the pancreas that regulates carbohydrate metabolism.

Heparin is an anticoagulant found in the liver and other tissues.

Sulfolipids, biochemicals in the brain, liver and kidneys and found in many enzymes.

Mucopolysaccharides, are sulphur-containing substances together with collagen, forms the glue that connects all body tissues. Sulfur bonds are essential structural features in all connective tissue, including glucosamine and chondroitin.

Mucopolysaccharides account for the structural strength of tissues, but they also help regulate the transfer of nutrients, gases, and other substances through the cell walls. The effectiveness of the mucous membranes at keeping out invading organisms, the ability of the gut to absorb nutrients while keeping out larger proteins, and elasticity of the blood vessels and skin depend on the amount and quality of mucopolysaccharides present.

The body manufactures mucopolysaccharides, but these substances not only decrease as we get older, but also deteriorate in quality. Supplemental doses of mucopolysaccharides have shown to be beneficial in boosting health and improving symptoms in many diseases such as arthritis, bursitis, respiratory disease, headaches (including migraine), ulcers, and allergies.

Mucopolysaccharides reduce inflammation, encourage healing, strengthen tissues, and stimulate the immune system. They reduce the tendency of the blood to clot, lower blood levels of cholesterol and fats, and increase synthesis of nucleic acids, DNA and RNA.

Glucosamine/Chondroitin

Glucosamine and Chondroitin are naturally occurring mucopolysaccharides in the human body. Glucosamine is necessary for the production of Chondroitin, and is able to converts into Chondroitin, the fibrous protein substance that bind water in the cartilage matrix, and is key to normal cartilage metabolism. Studies show that individuals with arthritis have abnormally low levels of Chondroitin.

Glucosamine is derived from glucose and the amino acid glutamine. Certain natural nutritional substances are advocated as beneficial and important for proper joint function. Researchers have become especially excited about the significant clinical effects of glucosamine sulfate. It is almost devoid of toxicity, and is suitable for long-term use. (Setnikar)

Dr. M.F. McCarty in *Medical Hypotheses* states in regard to the numerous double-blind studies conducted on oral glucosamine supplementation, *"Medical researcher and physicians in the U.S. have totally ignored this rational and safe therapeutic strategy. These and other safe nutritional measures supporting proteoglycan synthesis, may offer a practical means of preventing or postponing the onset of osteoarthritis in older people or athletes."* (McCarty)

The best-selling book, *The Arthritis Cure* by Drs. Throdosakis, Adderly and Fox introduced these two revolutionary substances to the public not long ago and go further into detail on their benefits. However, it is uncertain whether supplementation of both substances is necessary. This is because glucosamine is a precursor to chondroitin. Therefore, only glucosamine is needed.

What Sulfur Can Do For You!

Sulfur has been used for arthritis, gout, bronchitis and constipation and many other health problems for years. This was in the form of DMSO which smells bad, is hard to obtain, and is not approved for human use.

DMSO has been used for many years, but no one really knew (or cared) why it worked. People simply used it.

We simply cannot maintain the integrity of our tissues without proper nutritional support. Because sulfur is a crucial component of the tissues, hormones, vitamins, enzymes, antibodies, and antioxidants, etc., we simply cannot maintain good health without it.

Most of the research on sulfur to date has been under the direction of Dr. Stanley Jacob and Robert Herschler at Oregon Health Sciences Institute. Dr. Jacob has been examining the effects of DMSO and MSM for over 30 years.

Most of the information on human application stems from his research. He also holds several MSM patents.

Rheumatoid Arthritis

Sulfur is necessary for the formation of connective tissue. MSM has been investigated for the treatment of arthritis and other complications of joint inflammation. Scientists around the world investigating the relationship between MSM and arthritis have shown that sulfur concentration in arthritic cartilage is only about one-third the level compared to normal tissue. (Rizzo) In addition, the level of cystine, one of the sulfur-containing amino acids, in arthritic individuals is usually much lower than normal.The results of several studies showed that supplementation of MSM, significantly reduced joint degeneration and inflammation.

Researchers at Oregon Health Sciences Institute and Reagan State Science University have discovered that there is significant reduction in pain and inflammation associated with arthritis among patients using dietary supplements or a frequent topical application of MSM.

As a component of mucopolysaccharides such a chondroitin sulfate, dermatan sulfate and hyaluronic acid. (Sharif), the sulfur in MSM contributes to the production of "ground substance" which keeps connective tissue intact. It is necessary to help maintain flexibility and elasticity.

MSM enhances the structural integrity of these mucopolysacharides found in high concentration in connective tissue, including joint cartilage.

Researchers at Oregon Health Sciences University studied a strain of mice that were prone to the spontaneous development of joint lesions similar to those in rheumatoid arthritis (Morton). They found that animals that were fed a diet that included a 3% solution of MSM as drinking water from the age of two months until the age of five months suffered no degeneration of articular cartilage. In a control group of mice receiving regular tap water, 50% of the animals were found to have focal degeneration of articular cartilage.

Clinical evidence gathered by scientists at Reagan State Science University show there is significant relief of pain and stiffness along with reduced swelling and inflammation among those arthritic patients using MSM. Even muscle pain associated with multiple sclerosis responded very favorably to MSM.

Researchers have claimed that MSM in sufficient levels as a dietary supplement can help the following:

*** Improve joint flexibility**

*** Reduce stiffness and swelling**

*** Improve circulation** and cell vitality

*** Reduce pain** associated with systemic inflammatory disorders such as arthritis

*** Reduce scar tissue** which further aggravates the condition, increasing pain and decreasing mobility.

*** Break up calcium deposits.** Arthritis may be induced by the intake of too much calcium and by taking the wrong forms of calcium. Calcium can migrate to the soft tissues and form deposits. The tissues become calcified and the cells cease to function normally. MSM is able to rupture the weak (water) bonds of calcium in the synovial fluid. This is likely to also be true for kidney stones.

MSM can benefit all types of arthritis

This is an incredible leap for some 40% of the population who suffer from some form of arthritis - the most common are rheumatoid and osteoarthritis.

All forms of arthritis are characterized by pain, swelling, stiffness, and deformation of one or more joints. Joints of the knees, wrists, elbows, fingers, toes, hips, shoulders, neck and spine can be affected. Arthritis may appear suddenly or come on gradually. The pain may be sharp, burning, grinding, or dull. Moving the joint usually hurts, although sometimes there is only stiffness.

In **rheumatoid arthritis** the synovial membranes surrounding the lubricating fluid in the joints degenerate. Cartilage tissues and often, bone surfaces are destroyed. The body replaces this damaged tissue with scar tissue. This removes the space between the joints. Fusion of bones may occur further limiting movement. The entire body is affected instead of just one joint as in osteoarthritis. Rheumatoid arthritis causes stiffness, swelling, fatigue, anemia, weight loss, fever; and often crippling pain.

Osteoarthritis is a degenerative joint disease often accompanying aging. It involves deterioration of the cartilage at the ends of the bones. Smooth surfaces of cartilage become rough, resulting in friction. The tendons, ligaments, and muscles holding the joint together weaken, and the joint itself becomes deformed, painful, and stiff. However, unlike rheumatoid arthritis, there is little or no swelling. Osteoarthritis rarely develops before age 40. It afflicts approximately 15.8 million Americans. It typically runs in families, and occurs almost three times more often in women than men.

Dr. Jacob studied 24 individuals with symptomatic osteoarthritis treated with either 2,400 mg Motrin (NSAID) or 3,000 mg MSM. After one month, both groups noted equal improvements in pain and stiffness.

However, long term use of NSAID can have dangerous effects and provides no nutritional support to cartilage as MSM does. MSM is completely safe, with no side effects.

3 to 4 grams MSM appears to be a minimal daily dosage. Ten to 20 grams a day provides even more dramatic results.

PETS!!! Yes! Pets (dogs, cats, horses, etc.) often suffer from joint pain and can greatly benefit from supplementing MSM. Horse people have known about the benefits of MSM for a long time. Dosage depends on weight, but it's non-toxic so you don't have to worry so much.

In their water bowl add about 1,000 mg (often 1 capsule). Just open and dump out the contents into the bowl and stir, For larger pets (over 50 lbs) add two capsules or 2,000 mg.

For animals experiencing joint pain over 50 lbs, 2,000 mg. daily may be required. Open the capsules and add the contents to their food (if moist) or place intact capsules inside a treat like peanut butter or soft cheese.

For pets over 100 lbs, 3-4 capsules, 1,000 mg each

may be required. Benefits should be seen within a week!

Systemic lupus erythematosus is a form of arthritis which, for unknown reasons, the body produces antibodies against itself. Lupus is similar to rheumatoid arthritis as it results in painful and inflamed joints. Currently 131,000 Americans are afflicted; eight to nine times as many women as men have lupus.

Lupus may begin abruptly, or over several months or years. The disease is characterized by periodic bouts of fever, fatigue, and patches of raised, red, skin rashes. Lupus can also affect the heart, the lungs, the spleen, the blood and the gastrointestinal tract. Lupus is considered an autoimmune disease since most patients are found to develop anti-nuclear antibodies in their blood at some time during the illness.

According to Dr. J. M. Siegel, at Oregon Health Sciences University, experiments conducted on "lupus-prone" mice showed MSM protected the mice before and after the onset of the disease. Mice on a diet with 3% MSM in their water supply from age one month suffered lower death rates and liver damage than control groups drinking only tap water. After seven months, 30% of the control group had died, while none of the MSM mice had died. Also, when mice seven months old and already showing signs of advanced lupus were fed the MSM diet, 62% of the animals were still alive after nine months compared to 14% for the control group that received only tap water.

TMJ (temporal mandibular joint dysfunction) is similar to arthritis, where joint tissue degenerates, and scar tissue builds up interfering with proper movement. Individuals experience headaches,jaw pain and difficulty chewing food. TMJ is often associated with injury such as a car accident or grinding of the teeth. There are no human studies that I know of using MSM for TMJ, so I will tell you about Sally.

Sally, in her late 20's, was in a serious car accident about four years ago. She developed TMJ dysfunction on both sides and experienced severe headaches on a daily basis. The condition declined to the point where she could not work and was on a liquid diet because she could not chew food. NSAID no longer provided any relief at any dosage.

Wanting to avoid surgery, as a last result, she started taking 20 grams of MSM daily in distilled water. She took about 10 grams calcium ascorbate (buffered form of vitamin C), DHA (essential fatty acid to help with inflammation), and glucosamine.

After two weeks, she experienced some improvement, and after four weeks her headaches were almost gone. The last time I spoke with her she had cancelled her appointment with her surgeon. She was still taking about 10 grams of MSM a day, experiencing almost no headaches and was able to chew many foods again. She felt there had been an 80% improvement and that her symptoms continued to improve.

For years "arthritis" sufferers could do little more than take anti-inflammatory agents. Steroids have side effects and also cease to work effectively when taken continually. Aspirin often upsets the gastrointestinal tract.

Additional Helpful Supplements:

Vitamin C (Calcium Ascorbate)

Vitamin C (at levels above 1,000 mg per day) increases natural anti-inflammatory cortisone production, is helpful to the adrenal glands, and boosts immune function. Vitamin C is effective against free radicals, stimulates white blood cells and overall immune function. In the process of forming new cells, the body uses up vitamin C.

Herbs such as Burdock root, Ginger, Bromelain, etc.

Bromelain, in combination with vitamin C and biofiavonoids (such as rutin), can be more effective than aspirin and other nonsteroidal anti-inflammatory drugs for reducing pain and swelling. (Lotz-Winter)

Allergies

MSM fortifies the body's natural barriers against allergens. Oral MSM has alleviated the allergic response to pollen and to foods. MSM is as good as or better than the traditional antihistaminic preparations, but without the side effects.

Sulfur plays a major role in alleviating allergies and many forms of lung dysfunction through detoxification and elimination of free radicals. Research studies have shown that MSM supplementation has the ability to enhance lung function and control diverse allergy responses to pollen and foods in as little as a few days.

Asthma

MSM also strengthens the lungs against allergic responses. It may help regulate the fluid that covers the surface of the airways (airway surface fluid, ASF). This is a critical component of one of the first defense mechanisms of the lung against insult from microbes and other environmental agents. ASF collected from healthy airways contains much less Na and Cl (approximately 45% less) and much more K (around 600% more) than extracellular fluid or plasma, which shows that steep ion gradients exist across normal airway epithelia. These differences also show that ASF composition must be regulated and maintained by active electrolyte transport processes of airway epithelia, and that it is not merely the evaporated residue of isotonic secretions or extracellular fluid exudate. However, in patients with asthma or other sustained airway irritation, infection, or cystic fibrosis, ASF composition appears to become more isotonic with respect to plasma and much more hypotonic in patients with asthma. (Joris)

As an anti-inflammatory agent, MSM can do much to help the asthmatic individual. As I mentioned in the introduction I personally experienced a remarkable bronchial clearing after just three days (taking 20 grams per day) and subsequently maintained on a dose of 10 grams per day. I can testify to less use of inhaled proventil/albuteral, fewer nighttime awakenings and less sensitivity to irritants.

Cancer

Cancer cells are abnormal cells which cannot fulfill their normal functions in the body. These cells multiply uncontrollably to form tumors which invade neighboring tissues, robbing normal cells of essential nutrients. Cancer cells can spread and progressively create havoc throughout the entire body. Symptoms depend on the type and location of the cancer.

The key to fighting cancer is to prevent it. Abnormal cells which are potentially cancerous are continually produced throughout the body. It is the responsibility of the white blood cells to rid the body of these abnormal cells before they can multiply and cause harm. But, continued exposure to carcinogenic substances such as tobacco, pesticide residues, and chemicals found in certain drugs, etc., weakens the immune system. If the body is maintained in good health, not abused or overwhelmed with chemicals in foods or drugs, stress, obesity, nutrient deficiencies, then the body can fight off cancerous cells, keeping them in check. Through diet and supplementation we can provide the immune system all it needs to keep us healthy.

Several different studies have examined sulfur-bearing vegetables and herbs and found that they strongly inhibit the development of carcinogens in the body. Dr. Lee Wattenberg and his associates at the Department of Laboratory Medicine and Pathology at the University of Minnesota in Minneapolis produced two published reports. They stated that cruciferous vegetables (cabbage, brussels sprouts, cauliflower, alfalfa and broccoli) significantly inhibited the formation of chemical carcinogens. They provide the body vital protection against tumor formations. Alfalfa and cabbage in particular, along with turnips, stimulated aryl hydrocarbon hydroxylase activity, which resulted in a loss of active carcinogens from the body. (Ershoff, Potter)

German researcher, Dr. Joanna Budwig, used a combination of sulfur-rich amino acid foods and Omega-3 flax seed oil to treat various forms of cancer. The sulfur-amino acids eliminate toxins and free radicals.

Breast Cancer

Research done at the Ohio State University College of Medicine shows that oral MSM can protect rats against breast cancer. Rats specially bred to be susceptible to breast cancer when given certain carcinogenic compounds were fed a diet containing added MSM for eight days. Following this preliminary period, they were given oral doses of cancer-causing chemicals. The health of these animals was monitored for nearly a year and compared to a similar group of carcinogen-dosed rats that had not received the MSM diet. Although there was no statistical difference in the number of tumors developing in the two groups, the MSM diet rats developed their first tumors some 100 days later than the unprotected rats, and these tumors became cancerous some 130 days later than the controls. The average life expectancy of rats is two years. This would make 100 days the equivalent to about 10 years in the human life. (McCabe)

Colon Cancer

These same research workers from Ohio State University Medical College also studied the protection dietary MSM gives to rats injected with dimethylhydrazine, a compound that induces colon cancer. One group of rats received MSM as 1% solution in their drinking water throughout the time of the experiment. The control group received only tap water. One week after the start of the dietary regimen all rats were injected with the carcinogen. At two month intervals the rats were examined for tumors under anesthesia. Rats without any appearance of tumors were returned to the experiment. Again the number of bowel tumors occurring in the rats was statistically the same for treated and untreated rats over the entire nine months that the experiment was continued. However, the time of the appearance of the first bowel tumors was considerably longer in the MSM treated rats. The researchers concluded that MSM significantly lengthens the time of tumor onset compared to the controls. MSM should be further investigated as a chemopreventative agent for colon cancer. (O'Dwyer)

Studies show that there is a consistent decrease in tumors of the colon and rectum when there is frequent ingestion of sulphur-bearing vegetables like cabbage, Brussels sprouts, broccoli, garlic and onions. (Potter, Wattenberg)

Diabetes

Diabetes is a metabolic disorder involving inefficient production and/or use of insulin. Many diabetics lose their ability to absorb insulin and glucose as their cells become impermeable and resistant to insulin. So even if you can produce insulin (or are injecting it), it's not able to do its job.

MSM can help both of these situations as sulfur is needed to produce insulin and other vital components necessary for healthy carbohydrate metabolism such as thiamine and biotin. MSM, at about 10 grams per day, can restore normal blood sugar levels as cells become permeable and can help restore the pancreas functions to normal as blood sugar is absorbed by the cells.

Digestive Disorders

Digestive disorders include gastritis, heartburn, and indigestion, which are symptoms of abnormal digestion. These are characterized by acute or chronic abdominal discomfort, pain, irritation, bloating or gas, and are often accompanied by general malaise, headache, nausea, and sometimes vomiting.

MSM increases nutrition and improves the following problems:

Hyperacidity/Heartburn: Individuals who use antacids and histamine receptor antagonists can benefit from MSM because it reduces stomach acid and heartburn.

Constipation: Individuals with chronic constipation have had prompt and continued relief with daily, oral supplements of 500 mg. of MSM per day. These individuals should also increase water consumption.

Diverticulitis, Ulcerative colitis, Crohn's disease: MSM benefits these conditions by acting as a natural anti-inflammatory agent to hasten healing.

Parasites: Clinical testing suggests MSM has activity against a variety of parasites, such as giardia (associated with "travellers diarrhea"), trichomonads, candida albicans, and round worms. MSM apparently discourages these infections by competing for binding receptor sites at the mucus membrane surface in the intestinal tract.

One individual with confirmed Giardia, was given 500 mg. MSM three times a day for 14 days. By the eighth day he no longer had symptoms and neither of 2 specimens collected one week apart showed infection.

After 17 days, rats fed a 2% MSM solution showed complete clearance of pinworms. (Herschler, U.S. Patent # 4,616,039)

Note: The author gives her two dogs a water solution containing MSM to ward off parasitic infection and keep them healthy.

Gum Disease

Gum disease, also called periodontal disease or pyorrhea, is the inflammation and infection of the structures that support the teeth. It is a progressive disorder caused by the bacteria found in plaque.

Plaque is an invisible film which forms on the teeth. If not removed every day it will build up under the gum line, and the bacteria will thrive and multiply. Plaque calcifies (called tartar), and can be removed only by your dentist. The bacteria emit toxins that inflame and irritate the gum tissue so it pulls away from the teeth. Inflammation can spread to the bone underneath and cause deterioration and tooth loss.

Periodontal disease is the major cause of tooth loss among adults. It is estimated that by age 60, nearly 40% of the adult population requires false teeth as a result. The condition usually affects people over age 30 and becomes more prevalent with age.

Test subjects who had not had their teeth professionally cleaned for four to six months and were experiencing gum irritation were given a paste prepared by combining MSM on a 50/50 ratio. Subjects brushed their teeth twice daily. Following one week of use, all test subjects were free of signs of inflammation. (Herschler, U.S. Patent # 4,616,039)

One can also use MSM in water as a mouth rinse or gargle to help eliminate odors, and to maintain healthy teeth and gums.

Hair

Nutritional deficiencies can be one of the major causes of hair problems. Optimal blood circulation is dependent upon nutrition. A well-balanced diet is important for maintaining healthy hair, although hereditary graying and balding cannot be completely prevented by nutritional means.

Hair is 98% protein (mostly keratin). A deficiency of sulfur or any amino acids, (such as cystine), can result in color change, texture change, or hair loss. If the deficiency is corrected, the hair will return to its normal condition.

Nasal Congestion (Sinusitis)

Nasal congestion occurs when blood vessels in the nose enlarge, taking up space in the nasal cavity. This restricts the amount of airflow and prevents normal breathing.

Sinusitis is an acute inflammation of the accessory nasal sinuses, nasal congestion, and postnasal discharge. Sinusitis is often accompanied by headache, pain behind the eye, tenderness, fever, and loss of smell.

Sinuses are air-containing spaces located within the skull. The membranes within them secrete mucus that clean out the passages protecting us from infection.

MSM, at dosages of 1 to 3 grams daily strengthens the integrity of the mucous membrane tissues.

Pain and Inflammation

MSM has powerful anti-inflammatory, pain reducing properties. MSM blocks pain response in certain nerve fibers. It reduces scar tissue by altering cross-linkages which contribute to scar formation. Together these allow tissue repair and healing to take place.

How? Sulfur makes cells permeable by regulating the sodium potassium pump. This allows fluids and nutrients to flow freely through cell walls. This process removes toxins in the cell, reduces pain and inflammation and promotes healing. But every time the body removes invading toxins from the cell, it also removes the sulfur compound that neutralizes the toxins in the first place. This is why sulfur compound nutrients in the form of MSM are so vital to daily nutrition.

MSM is a fascinating, life- enhancing, anti-inflammatory agent for individuals with degenerative or rheumatoid arthritis, disc problems in their back, acute injuries, tendonitis, bursitis and other similar problems. Dr. Jacob claims it is certainly an important adjunct providing long-lasting relief when used on a continual basis.

MSM reduces lactic acid buildup and has the ability to reduce the incidence of, or entirely eliminate muscle, leg, and back cramps. This is especially beneficial for older individuals who experience cramps at night or after long periods of inactivity.

Among athletes, along with reducing leg cramps, the recovery time of cramping in marathon runners who were given MSM dropped by 75%. (Jacob)

Skin

Sulfur is called nature's "beauty mineral" because it keeps the hair glossy and smooth, and also keeps the complexion clear and youthful.

Sulfur is necessary for to produce collagen and keratin, protein substances necessary for health and maintenance of the skin, nails, and hair. Cystine makes up about 14% of skin and hair tissues. As a component of cystine, sulfur helps to heal and repair most tissues in the body. This includes those at risk from damage due to free radicals, aging, physical injury to internal organs and the skin. MSM is also important for recovery from burns and surgical incisions. Scar tissue will result without adequate cysteine and sulfur.

Because of its ability to protect against the harmful

effects of radiation and pollution, sulfur slows down the aging process and extends life span. It is found in hemoglobin and all body tissues and is utilized in the synthesis of collagen. This prevents dryness and maintains elastin in the skin.

Teenagers found that MSM (100 to 1,000 mg daily) experienced improvement in acne, including rosacea, in about one week of use. (Herschler, U.S. Patent # 4,616,039)

Itching and dry, scaly skin can also be improved with MSM at an oral dosage of between 100 and 1,000 mg daily in about a week.

MSM may be used oral or topically to aid shin disorders. When used topically in the form of an ointment or lotion, sulfur is helpful in treating skin disorders including acne, psoriasis, eczema, dermatitis, dandruff, scabies, diaper rash and certain fungal infections.

Burns

To reduce pain, hasten healing, and reduce scarring, soak burned area in a solution of MSM and water several times a day for at least 10 minutes. Increase daily oral intake to at least 2 to 10 grams daily, depending on the severity of the burn.

Snoring

Research at Oregon Health Sciences University on 35 individuals suffering from chronic snoring has shown that MSM as a 16% water solution administered to each nostril 15 minutes before sleep significantly reduced snoring in 80% of the subjects after one to four days of use.

In eight of the individuals experiencing relief with MSM, saline solution was substituted for MSM without their knowledge. Seven of eight individuals resumed loud snoring within 24 hours of the substitution. After the MSM treatment was restored, these eight people again showed a significant reduction of snoring. After 90 days of treatment none of the subjects reported any toxic reactions.

The MSM preparation has now been granted a patent (U.S. Patent 5,569,679) as an aid to the relief of snoring.

In addition to these there are hundreds of anecdotal stories from individuals who claim MSM has helped them with alzheimer's, cognitive function, stress, anxiety, and depression. (Mindell recommends using MSM starting at 9,000 mg daily and tapering down to 3,000 mg.)

How to Use MSM

MSM is available in capsules or as a powder. The daily dose may range from 2 to 20 grams, depending on your diet, body weight, state of health, etc. The precise dose is not really that critical as it is considered a food. For the first week or two larger doses may be needed to compensate for a deficiency.

Some forms of minerals, particularly inorganic minerals, are poorly absorbed. The organic sulfur in MSM is extremely well absorbed. Administered orally, a portion of MSM binds to the mucosal membrane receptor sites, while the rest is quickly absorbed into the bloodstream. Cellular membrane permeability allows MSM to readily penetrate cell membranes, as well as subcellular fractions, such as the mitochondrial, lysosomal and nuclear portions. Within 24 hours of administration, the sulfur derived from MSM can be found in the liver, kidney and blood.

"Therapeutic" Use of MSM

For those who have arthritis (or similar degenerative condition), asthma, digestive disorders (including parasites), or some other "serious" health problem, it is best to "detox" first using higher levels of MSM to obtain quickest results. Everyone is different, and different tissues in the body respond at different rates.

Lungs, considered one of the vital organs of the body, respond more quickly than arthritic joints. You could notice a difference in your lungs in as little as 3 days, while other conditions such as arthritis may take 2-3 weeks to see improvement. The longer it is used, the better the results. The higher the amount used, the quicker the results.

First Week: "Detox"

20-30 grams MSM per day.

One may wish to split this into 1/2 of the dosage in capsules and 1/2 of the dosage mixed into **distilled water**. Dissolve 1 teaspoon of MSM into 12 ounces of water. There is no taste. Drink **at least five 12 ounce glasses** of MSM water*, spread throughout the day before 6 pm**. This is a total of 18 grams MSM. Additional MSM can be supplemented through capsules, readily available in 750 mg. or 1 gram strengths.

Twice a day (first thing in the morning and again in the late afternoon), add one package Alacer's Emergen-C (available at health food stores) or similar product, to the water. This product provides highly soluble mineral ascorbates which are needed to heal and repair tissues. Especially important are potassium, zinc, manganese, and calcium ascorbate (Vitamin C). It provides 25 different electrolytes in all plus C and B vitamins.

* *Water is very important to help flush toxins out of the body. More water, the faster the results, and the fewer side effects will be experienced from the detoxification process.*

** *Some individuals find that MSM gives them extra energy so in order to avoid sleep difficulties, avoid taking any MSM after 6 pm.*

Possible side effects: One may experience headaches and fatigue for the first few days due to the detoxification process. If this occurs, drink more water. Do not be alarmed as this is a natural part of the healing process.

2nd Week

Continue until improvements in symptoms is experienced.

Take 10-20 grams MSM daily.

Add 1/2 teaspoon MSM per 12 ounces of water. Drink at least four 12 ounce glasses of MSM water, spread throughout the day before 6 pm. This is a total of 8 grams MSM. Additional MSM can be supplemented through capsules.

Twice daily, add one package Emergen-C to the water.

Therapeutic Maintenance

Take 2 to 10 grams MSM daily.

To simplify things, (and so the entire household can benefit from the restored sulfur diet), add MSM powder directly into a water dispenser. When it is time to switch jugs, drain several cups of water into a pitcher and dump in the desired amount of MSM. Stir until dissolved. It will take a few minutes. Dump this mixture into the new jug before placing it into the dispenser.

1 teaspoon = 4 grams MSM
1 gallon jug = 4.5 teaspoons MSM
5 gallon jug = 1/2 cup MSM (24 teaspoons)

How to Make "Maintenance" MSM Water

Add 1/2 teaspoons MSM for every 12 ounces of water. One should drink 64 ounces of water per day. This is about 10 grams MSM. At this level, it is tasteless and it's the most natural way to supplement. Extra capsules will not be needed.

In the morning, add one package Emergen-C to the water.

Although MSM is highly water soluble and it mixes quickly at room temperature water in a glass with a spoon, it takes a little longer with cold water. One may use a personal handheld mixer to speed things up.

Note: Some authorities believe that sulfur is more effective when taken with Vitamin C, B Complex vitamins, especially B-1, B-5 and biotin, and electrolytes. Be sure to drink adequate amounts of water daily.

Topical MSM

Lotions, creams or gels that contain MSM applied directly to the skin are a very effective ways to apply organic sulfur to an inflamed sore joints, muscles or to troubled and injured skin areas. A therapeutic quality of MSM allows the MSM to saturate the tissues directly and heal the joints, skin and muscular skeletal system.

There are several products on the market containing MSM. I have not compared them to see if one is better than the other, I would simply select a reputable manufacture that you can trust.

"Non-Therapeutic" Use of MSM

To maintain good health, regular "non-therapeutic" maintenance daily dosage range between 4 to 8 grams. One may reduce the recipe to the left to 1/2 strength if desired. Be sure to divide your daily dose in half (am and pm) as MSM is highly water soluble. If you want to take more than 8 grams a day, take 4,000 mg three times daily, instead of 6,000 twice, for example.

Safer Than Salt!

MSM is rated as one of the least toxic substances in biology, similar to water. Common table salt is much more toxic than MSM. The lethal dose of MSM for mice is over 20 grams per kilogram of body weight. By comparison, common table salt is toxic at 2.5 grams per kilogram of body weight.

When MSM was administered orally to human volunteers, no toxic effects were observed at levels up to one gm/kg of body weight per day for 30 days (Jacob). Intravenous injections of 0.5 gm/kg daily for five days a week produced no measurable toxicity in human subjects.

If one does take more MSM than needed, it simply passes through the body and is excreted in the urine. The excess MSM won't harm anyone.

MSM has been widely tested as a food ingredient without any reports of allergic reactions. An unpublished Oregon Health Sciences University study of the long-term toxicity of MSM over a period of six months indicated no toxic effects. More than 12,000 patients have been treated at the Oregon Health Sciences University with MSM at levels above two grams daily with no serious toxicity. (Jacob)

Dr. Jacob reports he personally takes 30 grams MSM per day, and has done so for 20 years. He reports he has not had a cold or flu since he began this regimen, which had previously experienced once or twice every year.

What Organic Sulfur is NOT

Do not confuse MSM with sulfa-based drugs, sulfites, or sulfates.

Sulfa-based drugs (sulfonamide) are part of a group of high molecular weight (man-made) compounds which have been known to cause allergic reactions. Sulfa drugs include erythromycin, sulfisoxazole, sulfacytine, sulfamethoxazole, and sulfasalazine. These are large complex molecules (drugs) used as antibiotics. It is highly unlikely that anyone would be allergic to MSM or sulfur, a naturally occurring substance in the body and many foods.

Sulfites are preservatives, antioxidants and browning agents used in foods. Ingestion of these is associated with adverse reactions such as asthma attacks, nausea, and diarrhea. There are several sulfiting agents now in use: sodium sulfite, sodium bisulfite, potassium bisulfite, sodium metasulfite and potassium metasulfite.

Sulfates are sulfuric acid salts.

Sulfuric Acid is a heavy corrosive oily acid used in producing fertilizers, chemicals and petroleum products.

The sulfa drugs that many people are allergic to is NOT the same as sulfur or MSM supplements!!!

Where Does Supplemental MSM Come From?

I have already explained how DMSO and MSM are made by nature in the clouds. MSM Supplements available today are produced synthetically, or may be produced by a combination of natural and synthetic processes.

MSM

comes from

DMSO

which comes from two different places

Di-methyl Sulfide
(used in petrochemicals, industry and natural gas industry)

Methanol
(Product of natural gas)
+ Hydrogen Sulfide

Synthetic

Elemental Sulfur
reacted with
Spent Kraft Ouloing Liquor.
This comes from trees where pulping has been done and the cellulose has been removed. DMSO is a natural by-product.

Natural and Synthetic

Shopping for Quality MSM

Be sure to look for 100% pure MSM. Some companies will add flow agents like cellulose and hemicellulose because MSM tends to clog up the encapsulation machines. Avoid these additive because they are fiber which binds to minerals and will pull much of the sulfur right out of the body. If you taking a MSM product containing additives you might have to take double the dosage to get the results you would get with a pure product! You may have to spend a little more to get a pure product, but it's worth it!

Bibliography

Agostini G; Martini P; Agostini S; Dellavalle F [Active properties and therapeutic effects of San Giovanni Spa mud Insegnamento di Idrologia Medica, Universita degli Studi-Pisa. Minerva Med 1996 Sep;87(9):427-32

Brown, Myrtle L., ed. Present Knowledge in Nutrition, International Life Sciences Institute, Nutrition Foundation, Washington, D.C. 1990, pg. 11

Ershoff, B.H., Protective effects of Dietary Fiber in rats ted toxic doses of docium cyclamate and polyoxyethylene sorbitan monsterate (Tween 60) Journal of Food Science 1975L 358.

Giordano N; Nardi P; et al *[The efficacy and safety of glucosamine sulfate in the treatment of gonarthritis]* Clin Ter 1996 Mar;147(3):99-105

Hegedus ZL; Nayak U *Homogentisic acid and structurally related compounds as intermediates in plasma soluble melanin formation and in tissue toxicities.* Arch Int Physiol Biochim Biophys 1994 May-Jun;102(3):175-81

Herschler, U.S. Patent # 4,559,329 (December 17,1985)

Herschler, U.S. Patent # 4,616,039 (October, 7, 1986)

Herschler, U.S. Patent # 5,569,679 (October 29, 1996)

Inaba M; Yukioka K; et al *Positive correlation between levels of IL-1 or IL-2 and 1,25(OH)2D/25-OH-D ratio in synovial fluid of patients with rheumatoid arthritis.* Life Sci 1997; 61(10):977-85

Jacob, S.W. and Herschler, R., Ann N. Y. Acad Sci, Vol. 411, xiii, 1983

Herschler, U.S. Patent 5,569,679 (October 29, 1996)

Janssen M; Dijkmans BA; Lamers CB *Upper gastrointestinal manifestations in rheumatoid arthritis patients: intrinsic or extrinsic pathogenesis?* Scand J Gastroenterol Suppl 1990;178:79-84

Joris L; Dab I; Quinton PM Elemental composition of human airway surface fluid in healthy and diseased airways. Am Rev Respir Dis 1993 Dec;148(6 Pt 1):1633-7

Kim, Organic Sulfur Chemistry: Biochemical Aspects, Ed Shigeru Oae and Tadashi Okuyama, CRC Press, Boca Raton, FL, 1992,138-140.

Kremer JM; Bigaouette J *Nutrient intake of patients with rheumatoid arthritis is deficient in pyridoxine, zinc, copper, and magnesium.* J Rheumatol 1996 Jun;23(6):990-4

Lakshmi R; et al *Effect of riboflavin or pyridoxine deficiency on inflammatory response.* Indian J Biochem Biophys 1991 Oct-Dec;28(5-6):481-4

Larsen NE; Lombard KM; Parent EG; Balazs EA Effect of hylan on cartilage and chondrocyte cultures. Department of Biochemistry, Matrix Biology Institute, Ridgefield, NJ 07657. J Orthop Res 1992 Jan;10(1):23-32

Lotz-Winter H *On the pharmacology of bromelain: an update with special regard to animal studies on dose-dependent effects.* Planta Med 1990 Jun;56(3):249-53

Lovelock, J.E. et al., Nature, Vol. 237, 452, 1972

Malfait AM; Verbruggen G; et al *Coculture of human articular chondrocytes with peripheral blood mononuclear cells as a model to study cytokine-mediated interactions between inflammatory cells and target cells in the rheumatoid joint.* In Vitro Cell Dev Biol Anim 1994 Nov;30A (11): 747-52

McCabe, D. et al., Arch Surg, Vol. 121,1455,1986

McCarty MF *The neglect of glucosamine as a treatment for osteoarthritis--a personal perspective.* Med Hypotheses 1994 May;42(5):323-7

McCully, Kilmer, M.D. The Homocysteine Revolution, Keats Publishing, 1997,

Miyazaki H; Sakao S; Katoh Y; Takehara T Correlation between volatile sulphur compounds and certain oral health measurements in the general population. J Periodontol 1995 Aug;66(8):679-84

Morton, J.I. & Moore, R.D., Federation of American Societies for Experimental Biology, 69th Annual Meeting, April 21-26, 1985, 692, 1985

Newnham RE *Agricultural practices affect arthritis.* Nutr Health 1991;7(2):89-100

Nielsen FH *Biochemical and physiologic consequences of boron deprivation in humans.* Environ Health Perspect 1994 Nov;102 Suppl 7:59-63

Oae, S., Okuyama, T., Organic Sulfur Chemistry: Biochemical Aspects, CRC Press, Boca Raton, FL.,1992.

O'Dwyer, P.J. et al., Cancer; Vol. 62, 944, 1988

Orkin M; Maibach HI Scabies therapy--1993. Department of Dermatology, University of Minnesota, Robbinsdale. Semin Dermatol 1993 Mar;12(1):22-5

Potter JD Steinmetz K, Vegetables, fruit and phytoestrogens as preventive agents. IARC Sci Publ 1996;(139):61-90

Rayssiguier Y; *Magnesium and ageing.* I. Experimental data: importance of oxidative damage. Magnes Res 1993 Dec;6(4):369-78

Reichelt A; Forster KK; Fischer M; et al *Efficacy and safety of intramuscular glucosamine sulfate in osteoarthritis of the knee. A randomised, placebo-controlled, double-blind study.* Arzneimittelforschung 1994 Jan;44(1):75-80

Rezvukhin AI; Krysin AP; Shalaurova IIu [Stabilization of omega-3-polyunsaturated acids in fat from Mallotus villosus (Osmeridae) by the nontoxic, sulfur-containing antioxidant CO-3] Vopr Med Khim 1995 May-Jun;41(3):37-9

Rizzo R; Grandolfo M; et al *Calcium, sulfur, and zinc distribution in normal and arthritic articular equine cartilage: a synchrotron radiation-induced X-ray emission (SRIXE) study.* J Exp Zool 1995 Sep 1;273(1):82-6

Roubenoff R; Roubenoff RA; et al *Abnormal vitamin B6 status in rheumatoid cachexia. Association with spontaneous tumor necrosis factor alpha production and markers of inflammation.* Arthritis Rheum 1995 Jan;38(1):105-9

Schauss, Alexander, Diet. Crime and Delia quency, Parker House, Berkeley, 1981.

Sharif M; Osborne DJ; et al *The relevance of chondroitin and keratan sulphate markers in normal and arthritic synovial fluid.* Br J Rheumatol 1996 Oct;35(10):951-7

Schiller J; Arnhold J; Sonntag K; Arnold K NMR studies on human, pathologically changed synovial fluids: role of hypochlorous acid. Magn Reson Med 1996 Jun;35(6):848-53

Stoner GD; Mukhtar H *Polyphenols as cancer chemopreventive agents.* J Cell Biochem Suppl 1995;22:169-80

Suarez Fernandez R; et al Norwegian scabies in a patient with AIDS: report of a case. Department of Dermatology, Cutis 1995 Jul;56(1):57-60

U.S. Patent 5,569,679 (October 29, 1996)

Vellini M; Desideri D; et al *Possible involvement of eicosanoids in the pharmacological action of bromelain.* Arzneimittelforschung 1986;36(1):110-2

Lee W. Wattenberg, et al., "Dietary constituents altering the responses to chemical carcinogens," Federation Proceedings 38 (May 1976): 1327-31.

Lee W. Wattenberg and William D. Loub, "Inhibition of Polycyclic Aromatic Hydrocarbon-induced Neoplasia by Naturally Occurring Indoles," Cancer Research 38 (May 1978), 1410-13.

Williams, K.l.H. et al., Arch Biochem Biophys, Vol. 113, 251,1966

Zentner A; Heaney TG An in vitro investigation of the role of high molecular weight human salivary sulphated glycoprotein in periodontal wound healing. J Periodontol 1995 Nov;66(11):944-55

# of copies

TO PLACE AN ORDER:

___ *Aspirin Alternatives: The Top Natural Pain-Relievers* (Lombardi) . . . $8.95

___ *Bilberry & Lutein: The Vision Enhancers!* (Ley) . . . $4.95

___ *Calcium: The Facts, Fossilized Coral* (Ley) . . . $4.95

___ *Castor Oil: Its Healing Properties* (Ley) . . . $4.95

___ *Dr. John Willard on Catalyst Altered Water* (Ley) . . . $5.95

___ *Chlorella: Ultimate Green Food (Ley)* . . . $5.95

___ *CoQ10: All-Around Nutrient for All-Around Health* (Ley) . . . $4.95

___ *Coleus Forskohlii: Metabolic Modifier- Shape Up & Slim Down (Ley)* . . . *$4.95*

___ *Colostrum: Nature's Gift to the Immune System* (Ley) . . . $5.95

___ *DHA: The Magnificent Marine Oil* (Ley) . . . $6.95

___ *DHEA: Unlocking the Secrets/Fountain of Youth-2nd ed.* (Ash & Ley) . . . $14.95

___ *Diabetes to Wholeness* (Ley) . . . $9.95

___ *Discover the Beta Glucan Secret* (Ley) . . . $3.95

___ *Fading: One family's journey ... Alzheimer's* (Kraft) . . . $12.95

___ *Flax! Fabulous Flax!* (Ley) . . . $5.95

___ *Flax Lignans: Fifty Years to Harvest* (Sönju & Ley) . . . $4.95

___ *God Wants You Well* (Ley) . . . $14.95

___ *How Did We Get So Fat? 2nd Edition* (Susser & Ley) . . . $8.95

___ *How to Fight Osteoporosis and Win!* (Ley) . . . $6.95

___ *Maca: Adaptogen and Hormone Balancer (Ley)* . . . $4.95

___ *Marvelous Memory Boosters* (Ley) . . . $3.95

___ *Medicinal Mushrooms: Agaricus Blazei Murill (Ley)* . . . $4.95

___ *MSM: On Our Way Back to Health W/ Sulfur* (Ley) SPANISH . . . $3.95

___ *MSM: On Our Way Back to Health W/ Sulfur* (Ley) . . . $4.95

___ *Natural Healing Handbook* (Ley) . . . $19.95

___ *Nature's Road to Recovery: Nutritional Supplements for the Alcoholic & Chemical Dependent* (Ley) . . . $5.95

___ *pH: Optimal Health Through Alkaline Foods* . . . *$5.95*

___ *PhytoNutrients: Medicinal Nutrients in Foods*, *Revised /Updated* (Ley) . . . $5.95

___ *Recipes For Life! (Spiral Bound Cookbook)* (Ley) . . . $19.95

___ *Secrets the Oil Companies Don't Want You to Know (LaPointe)* . . . $10.00

___ *Spewed! How to Cast Out Lukewarm Christianity through Fasting and a Fasted Lifestyle -* . . . $15.95

___ *The Potato Antioxidant: Alpha Lipoic Acid* (Ley) . . . $6.95

___ *Vinpocetine: Revitalize Your Brain w/ Periwinkle Extract!* (Ley) . . . $5.95

Subtotal $ ________ Please add $5.00 for shipping. **TOTAL $** ________

Credit card orders call toll free: 1-877-BOOKS11

Order at: www.blpublications.com